AF270469

WILD STEM

STEM in the SKIES

Megan Borgert-Spaniol

Checkerboard
Library

An Imprint of Abdo Publishing
abdobooks.com

abdobooks.com

Published by Abdo Publishing, a division of ABDO, PO Box 398166, Minneapolis, Minnesota 55439. Copyright © 2024 by Abdo Consulting Group, Inc. International copyrights reserved in all countries. No part of this book may be reproduced in any form without written permission from the publisher. Checkerboard Library™ is a trademark and logo of Abdo Publishing.

Printed in the United States of America, North Mankato, Minnesota
102023
012024

Design: Denise Hamernik, Mighty Media, Inc.
Production: Mighty Media, Inc.
Editor: Anna Anderhagen
Cover Photographs: Imgorthand/Getty Images
Interior Photographs: Aerovista Luchtfotografie/Shutterstock Images, p. 5; Artsiom P/Shutterstock Images, p. 25; AstroStar/Shutterstock Images, p. 13; dima_zel/iStockphoto, p. 15; Frame Stock Footage/Shutterstock Images, p. 7; gorodenkoff/iStockphoto, p. 17; Juice Flair/Shutterstock Images, p. 19; Mighty Media, Inc., pp. 10–11, 20–21; milehightraveler/iStockphoto, p. 9; Supamotionstock.com/Shutterstock Images, p. 27; Triff/Shutterstock Images, p. 29; zieusin/Shutterstock Images, p. 23
Design Elements: Lorthois Yuliya/Shutterstock Images; supanut/Adobe Stock

Library of Congress Control Number: 2023939335

Publisher's Cataloging-in-Publication Data
Names: Borgert-Spaniol, Megan, author.
Title: STEM in the skies / by Megan Borgert-Spaniol
Description: Minneapolis, Minnesota : Abdo Publishing, 2024 | Series: Wild STEM | Includes online resources and index.
Identifiers: ISBN 9781098291983 (lib. bdg.) | ISBN 9781098278885 (ebook)
Subjects: LCSH: Science--Study and teaching--Juvenile literature. | Technology--Study and teaching--Juvenile literature. | Engineering--Study and teaching--Juvenile literature. | Mathematics--Study and teaching--Juvenile literature. | Sky--Juvenile literature.
Classification: DDC 577.0--dc23

CONTENTS

WILD SKIES

Do you love watching clouds and aircraft overhead? Do you dream of flying through Earth's atmosphere or even beyond it? If so, you might enjoy a career that takes you into the sky!

Many scientists study the sky. Meteorologists research and track weather in Earth's atmosphere. Aerospace engineers design aircraft and spacecraft. Pilots fly vehicles that aerospace engineers design. Astronomers study outer space, while astronauts travel there!

What kind of scientist would you like to be? There's only one way to find out. Buckle up and prepare for liftoff. It's time to explore science, technology, engineering, and math (STEM) in the skies!

STEM

Science
Technology
Engineering
Math

Clouds may look light and fluffy, but they are heavy. The water droplets in clouds can make them weigh more than one million pounds (450,000 kg)!

METEOROLOGISTS

What do you see in the sky today? Is it sunny with a few puffy clouds? Perhaps the sky is filled with wind, rain, and lightning. These are all examples of weather!

Events in Earth's atmosphere cause weather. For example, the sun's heat warms air in the atmosphere, causing air to rise. As warm air rises, cold air flows in to replace it. This movement of air creates wind. The wind changes the weather, such as by turning sunny skies into thunderstorms.

Weather **impacts** everything from daily clothing choices to food crops to power grids. Scientists study atmospheric events to better understand weather and forecast changing conditions. These scientists are called meteorologists.

Meteorologists predict the severity of storms. Knowing weather conditions in advance can save lives!

A big part of a meteorologist's job is collecting data. Meteorologists **monitor** Earth's surface temperature, precipitation, and wind speed from the ground. But data collection also takes place in the sky. For example, meteorologists release weather balloons. As these balloons rise through the atmosphere, they measure air pressure, temperature, and **humidity**.

Meteorologists even gather weather data from outer space! Weather **satellites orbit** Earth, capturing images of atmospheric conditions. These images help scientists track daily weather around the world. The images also allow scientists to monitor tornadoes and hurricanes in specific locations. This technology helps meteorologists forecast the weather!

Meteorologists prepare a weather balloon for an air quality project. Weather balloons can rise to an altitude of more than 18 miles (29 km) before bursting!

GET WILD!

BALLOON BAROMETER

Air pressure is the force of air molecules pressing on Earth. A barometer measures this force. Learn how temperature affects air pressure with your own barometer!

WHAT YOU NEED:

- ✔ balloon
- ✔ scissors
- ✔ glass jar
- ✔ rubber band
- ✔ drinking straw
- ✔ tape
- ✔ sheet of card stock paper
- ✔ marker
- ✔ 2 bowls
- ✔ hot water
- ✔ ice water

WHAT YOU DO:

1. Cut the neck off of the balloon. Stretch the balloon over the opening of the jar. Secure it with a rubber band.

2. Tape the end of the straw to the middle of the stretched balloon. The straw should extend beyond the jar's rim.

3. Fold a piece of card stock paper in half lengthwise. Stand the folded paper up next to the jar. Draw a line on the paper where the straw is pointing.

4. Place the jar in a bowl of hot water for one minute. Then move the jar back to its place by the folded paper. Draw a new line on the paper where the straw is pointing and label it "hot."

5. Place the jar in a bowl of ice water for one minute. Then move the jar back to its place by the folded paper. Draw a new line on the paper where the straw is pointing and label it "cold."

6. Compare the lines you drew. What happened to the barometer when it was placed in hot water compared to cold water?

ASTRONOMERS

While some scientists study Earth's atmosphere, others set their sights beyond it. Astronomers are scientists who study outer space. Some astronomers focus their research on planets. Others study the stars. And some search for clues about the history of the entire universe!

Astronomers study objects that are anywhere from thousands of miles to billions of **light-years** away. They need powerful instruments to see these objects. One important astronomy tool is the telescope.

Most telescopes use mirrors to focus visible light from the night sky. This focused light allows scientists to see distant space objects. Other telescopes use radio waves and X-rays to collect light waves that humans cannot see. These invisible waves inform scientists about a space object's structure and motion.

Astronomers work during the night and day. Some telescopes are so strong that astronomers can see objects in the sky even during daylight hours.

Scientists use telescopes in land-based **observatories**. But some telescopes operate from outer space! The Hubble and James Webb telescopes are space-based observatories run by the National Aeronautics and Space Administration (NASA). They pick up light from stars that are about 13 billion **light-years** from Earth!

Space probes are another tool that astronomers launch into space. Space probes collect images and send them back to Earth using radio waves. NASA's *Voyager 1* probe launched in 1977. It discovered five new moons **orbiting** Saturn.

While probes fly through space, rovers explore surfaces. Recently, rovers have traveled the dusty surface of Mars. Their explorations pave the way for sending humans to Mars in the future!

More than 20 space rocks have hit the James Webb Space Telescope. The 18 gold-plated mirrors are engineered to withstand direct hits from space rocks.

AEROSPACE ENGINEERS

Humans couldn't explore outer space without the help of aerospace engineers. These scientists design, test, and manufacture vehicles that move through Earth's atmosphere and outer space. Aeronautics and astronautics are the two main branches of aerospace engineering.

Aeronautics focuses on the flight of aircraft through Earth's atmosphere. These aircraft include **commercial** airplanes, helicopters, and military fighter jets. Astronautics focuses on the flight of spacecraft through outer space. These spacecraft include rockets, **satellites**, and probes.

The development of an aerospace vehicle begins with design. Engineers use computer **software** to create

Aerospace engineers build a satellite from a digital 3D model. This satellite will be part of a space exploration mission.

digital 3D models. Then, they turn these digital models into physical models. Engineers test the physical models to see how they will perform during flight.

Testing aircraft and spacecraft reveals any problems or weaknesses. Engineers figure out how to solve these problems with design adjustments. Then, they retest updated models until the vehicle meets their standards.

The work of aerospace engineers allows humans to travel the world, explore space, and even defend our planet. In 2021, NASA launched a spacecraft with the mission to **collide** with an asteroid millions of miles away. This mission tested whether humans could stop an asteroid headed toward Earth. The mission was a success!

Aerospace engineers work together assembling the turbine engine on a passenger jet to be flight-tested.

GET WILD!

STOMP ROCKET

Aerospace engineers design aircraft and spacecraft to move smoothly through the air. Design and launch your own rocket powered by air!

WHAT YOU NEED:

- ✔ 2-liter (67.6 oz) plastic bottle
- ✔ 2 feet (0.6 m) of plastic tubing, 1 inch (2.5 cm) wide
- ✔ duct tape
- ✔ sheet of paper
- ✔ tape
- ✔ scrap paper
- ✔ scissors

WHAT YOU DO:

1. Take the cap off the plastic bottle. Fit the plastic tubing to the bottle opening. Use duct tape to seal the tubing tightly to the bottle.

2. To make a rocket, wrap the sheet of paper tightly around the other end of the tubing. Tape the rocket body so it holds its shape.

3. Use scrap paper to create a nose cone and fins for the rocket. Tape these pieces to the body of the rocket.

4. Place the plastic bottle on the ground. Slide the bottom of the rocket onto the open end of the plastic tubing.

5. Aim the tubing away from you and anyone nearby. Then use your foot to stomp on the bottle.

6. How well does your rocket fly? Can you adjust the nose to make the rocket fly farther? Can you change the fins to make it fly straighter? What happens if you use heavier paper for your rocket body? Keep experimenting to perfect your rocket!

PILOTS

Most aircraft created by aerospace engineers need humans to operate them. Professionals who fly aircraft are called pilots. Many pilots fly **commercial** airplanes that transport passengers traveling for business or vacation. Other pilots fly military aircraft in combat or **reconnaissance** missions.

Pilots must be well trained from takeoff to landing. Flight schools use **simulators** to help with training. Flight simulators look like full-size cockpits and mimic the movement of an airplane. Simulators allow pilots to practice responding to different weather conditions.

Flying military aircraft requires intense training. For example, the U-2 spy plane is so difficult to land that a car follows it down the runway. The car driver radios the U-2's **altitude** so the aircraft pilot knows how close the plane is to the ground!

Every day, pilots fly millions of passengers around the world. At any given time, there are more than 8,000 airplanes in the sky!

ASTRONAUTS

While some pilots stay within Earth's atmosphere, others soar higher. People who fly to outer space in spacecraft are called astronauts. Pilot astronauts guide spacecraft. Mission **specialist** astronauts maintain equipment and conduct experiments.

Astronauts landed on the moon in 1969. And from 1998 to 2011, astronauts built the International Space Station (ISS). An international crew of seven astronauts conducts experiments and maintains the space station.

Before astronauts launch into space, they must train vigorously. This training helps astronauts prepare for **microgravity**. This small amount of gravity in space causes astronauts to float and feel weightless. During training, they practice performing tasks in conditions that **simulate** microgravity.

The ISS orbits Earth every 90 minutes. In 24 hours, astronauts see 16 sunrises and sunsets!

One of NASA's most important **microgravity simulators** is a giant swimming pool. At 40 feet (12 m) deep and 200 feet (61 m) long, the pool holds model parts of the ISS. Astronauts enter the pool wearing space suits and practice their mission tasks underwater. Practicing underwater prepares them for space walks. Space walks are trips outside the spacecraft to perform experiments or maintenance.

While doing space walks, bits of space dust zoom by at thousands of miles per hour. And there is no oxygen to breathe! Space suits protect astronauts from these deadly conditions. Astronauts know they are risking their lives with every trip past Earth's atmosphere. But little can stop these explorers from discovering new frontiers in space!

A space suit weighs about 280 pounds (127 kg). It takes an astronaut 45 minutes to put it on!

A CAREER IN THE SKIES

Do you have a passion for STEM in the skies? There are many career paths to choose from. Each path leads to a wide range of opportunities.

A meteorologist could work for a local TV station or government weather service. An astronomer might conduct research at a university or **observatory**. An aerospace engineer might work for NASA or an aircraft manufacturer. No matter where they work, these scientists share a love of the vast skies and space surrounding our planet.

Maybe one day, the sky will be your office!
What will you discover?

GLOSSARY

altitude—the vertical elevation of an object above a surface (such as sea level or land).

collide—to come together with force.

commercial—having to do with business or making money.

humidity—the amount of moisture in the air.

impact—to have a strong effect on something or someone.

light-year—the distance that light travels in one year.

microgravity—a condition of near weightlessness experienced in space.

monitor—to watch, keep track of, or oversee.

observatory—a place or a building for observing the weather or the stars.

orbit—to revolve in a circle around.

reconnaissance—military observation of the enemy.

satellite—a manufactured object that orbits Earth. It relays scientific information back to Earth.

simulate—to imitate. A simulator is a device that imitates a real experience.

software—the written programs used to operate a computer.

specialist—someone who pursues one branch of study.

ONLINE RESOURCES

To learn more about STEM in the skies, visit **abdobooklinks.com**. These links are routinely monitored and updated to provide the most current information available.

INDEX